MW00907359

CANADIAN BIRDS

CANADIAN BIRDS

Bruce Obee

Whitecap Books
Vancouver / Toronto

Edited by Elaine Jones
Cover design by Steve Penner
Interior design by Margaret Ng

Cover photograph by Robert Lankinen/First Light
Typography by CompuType, Vancouver, B.C.

Printed and bound in Canada by D.W. Friesen and Sons Ltd.

Canadian Cataloguing in Publication Data
Obee, Bruce, 1951-
Canadian birds

ISBN 1-55110-097-5
1. Birds—Canada—Pictorial works. I. Title.
QL685.023 1993 598.2971'022'2 C93-091536-4

Title page: Two young barn swallows (Hirunda rustica) *survey their new-found world from a fence picket. After breeding in Canada some migrate as far as South America.*

Contents

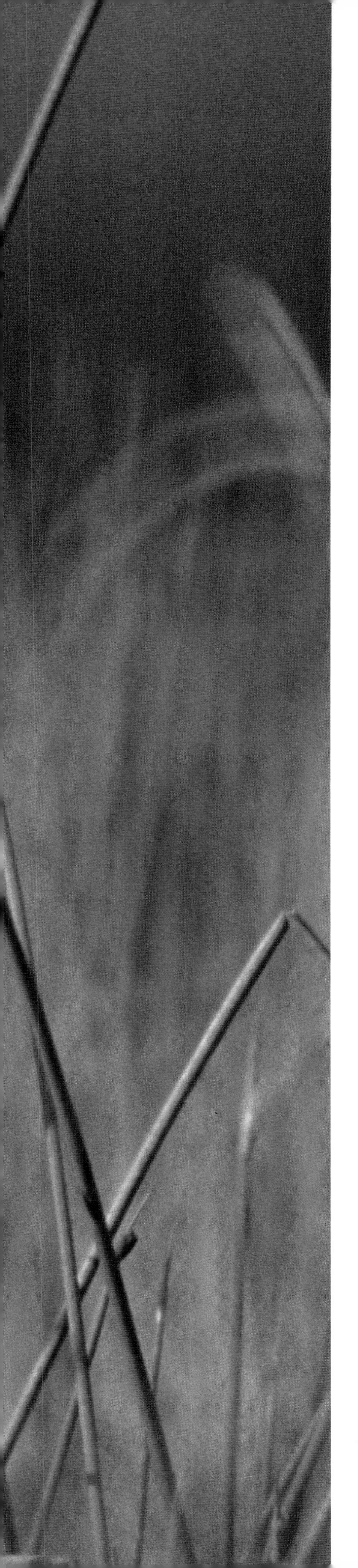

INTRODUCTION

Canada has an ecological niche for nearly every type of North American bird. Forests and tundra, peatland, mountains, rivers, lakes and wetlands, prairies and oceans all lie within Canada's boundaries. Nearly six hundred species occur here, in the world's second-largest country, and well over four hundred breed here. There's hardly a corner of Canada where birds cannot be found at all times of the year.

Where there are birds there are birders: it is estimated that as many as eight million Canadians feed and watch wild birds. Eighty-five percent of Canadians say they would willingly spend more to view birds and other wildlife. Birding today, next to gardening, is the most popular and fastest-growing outdoor pastime on the continent.

Birders play a significant role in the survival of birds through unforgiving Canadian winters. In some areas one

Like the red-winged blackbird, the yellow-headed blackbird (Xanthocephalus xanthocephalus) *builds basketlike nests over water in marsh reeds. Made of wet vegetation, the nests shrink as they dry, pulling the supporting reeds close together. Though they may nest in the same areas as their red-winged relatives, they generally construct their nests over deeper water, using similar but separate habitat.*

in three households sets out an average of thirty kilograms of birdseed a year. This boost from humans is helping some species expand their ranges. Birders who keep records and participate in seasonal bird counts also contribute immensely to our growing knowledge of bird habits and habitats. More and more, bird conservation is moving beyond the realm of government professionals to individual birders—people who join clubs and societies, who landscape yards with native plants, use fewer pesticides, and work with neighbours to preserve and enhance local bird habitats.

Volunteers concerned with the welfare of our avian fauna are the mainstay of such organizations as the Canadian Nature Federation, the World Wildlife Fund Canada, the Canadian Wildlife Federation, the Nature Conservancy of Canada and Ducks Unlimited Canada. These conservationists work closely with governments, land owners, industrialists, and others to protect and improve critical bird habitats across the country.

Canada was first in North America to establish an official bird sanctuary. Today the refuge at Last Mountain Lake, Saskatchewan, is as important to waterfowl and other birds as it was in 1887. And now there are a hundred-odd similar sanctuaries across the country, along with thousands of provincial and local parks and reserves. Federal sanctuaries are the responsibility of the Canadian Wildlife Service, which was designated in 1947 to administer the Canada-U.S. Migratory Birds Convention Act of 1917 in consultation with the ten provinces and two territories. The well-being of various species is monitored by the Committee on the Status of Endangered Wildlife in Canada.

Canada also works with other countries to protect its species and their environment. The nation is a signatory to the Convention on International Trade in Endangered Species. Under the international Ramsar Convention it is working toward preservation of crucial wetlands. Through the North American Waterfowl Management Plan, both Canada and the United States hope to

A wood duck drake (Aix sponsa) *peers out from its nesting cavity. Its resplendent plumage is one of nature's masterpieces: iridescent green, violet, purple, and bronze; stripes; speckles; and a colourful, flowing crest all immaculately arranged in one living package. Breeding wood ducks adapt well to artificial nest boxes, which are helping improve Canadian populations.*

enhance 1.5 million hectares of waterfowl habitat by the turn of the century.

Since the 1920s Canada and the United States have been sharing information derived from bird-banding programs: migration routes and timing, wintering areas, population dynamics, longevity, and breeding success. More than 40 million North American birds wear metal bands with serial numbers.

As modern-day pressures on vital bird habitats intensify, international diplomacy and negotiation become increasingly important. Canadian birds are long-distance travellers: there's not much point in preserving Arctic breeding grounds when those nesters of the north ingest DDT—restricted long ago in Canada—while feeding in Central and South American wintering areas. There's little value in enhancing Canadian staging areas if the next stopover along the southbound migration route is covered with condominiums. And Americans, who have banned lead shot for waterfowl hunting, are wasting much of their energy protecting waterfowl wintering areas if those birds come to Canada only to be poisoned by lead shot on Canadian fields and wetlands.

The value of North American birds both within and outside Canada now is recognized across the continent. It is the strength of that recognition that will determine the future of the birds of Canada.

The mournful mewing of the mew gull (Larus canus) *at its nest is the basis of its name. Known also as the short-billed gull, this seabird is a proficient fisher, taking herring and other small fish from near-shore waters. It is also a resourceful bird, hunting for insects and earthworms on freshly ploughed fields.*

SEABIRDS AND SHOREBIRDS

With more than 6.5 million square kilometres of ocean and 243,000 kilometres of coastline, Canada is home to a multitude of seabirds and shorebirds. Albatrosses, shearwaters, fulmars, and storm-petrels patrol offshore seas; cormorants, puffins, murres, guillemots, murrelets, auks, and razorbills nest in crowded colonies along Pacific, Atlantic, and Arctic shores. Islands and islets or inland sites across the continent are nesting grounds for gulls, terns, jaegers, kittiwakes, and skuas. Estuaries, shallow lagoons, mud flats, and sandy shores are foraging territories for herons, sandpipers, plovers, and other shorebirds.

Many of the country's busiest nesting and feeding sites are protected within parks or wildlife sanctuaries. Ecological reserves along the coast of British Columbia ensure secure habitats on

Northern gannets (Sula bassanus) *share a roost at Newfoundland's Cape St. Mary's Bird Sanctuary, the second-largest gannet colony in Canada. The most populated colony is Bonaventure Island, a five-square-kilometre provincial park in the Gulf of St. Lawrence. Each year more than forty thousand pairs of gannets gather on the cliffs at Bonaventure, attracting boatloads of ornithologists and tourists. These big seabirds help fishermen locate fish by diving from heights of twenty or thirty metres to splash through the surface and take their prey.*

Vancouver Island, the Queen Charlottes, and numerous smaller islands. Among them is Triangle Island, the province's largest seabird colony with more than a million nesting birds—tufted puffins, murres, auklets, pigeon guillemots, and more. Farther south the expansive beaches and tidal flats of Vancouver Island's Pacific Rim National Park are staging grounds for migrating shorebirds and waterfowl. On the Lower Mainland, the world's entire population of western sandpipers—1.5 million—stops at George C. Reifel Migratory Bird Sanctuary and other protected wetlands within the Fraser delta.

Across the country, Newfoundland's Witless Bay Bird Sanctuary is the centre of the North American breeding range for Atlantic puffins: more than 225,000 breed on islands in the bay. The sanctuary's Gull and Great islands are also nesting areas for 1.3 million Leach's storm-petrels and at least 100,000 murres. In Nova Scotia, Margaree Island National Wildlife Area is a roost for black guillemots, common terns, great cormorants, and black-backed gulls. Nearby Fundy National Park, in New Brunswick, is a 207-square-kilometre refuge for great blue herons and flocks of migrating shorebirds.

In the Northwest Territories, Polar Bear Pass, on Bathurst Island, was set aside in 1986 as a protected breeding ground for jaegers and several shorebird species. And under the international Ramsar Convention, Canada has agreed to conserve wetlands on Queen Maud Gulf, a significant waterbird breeding area on the shores of the northern mainland.

Every spring millions of seabirds converge on coastal colonies to mate and raise their offspring. Although sharing these isolated islets affords protection from predators (safety in numbers) nesting sites nonetheless are hunting grounds for raptors, racoons, otters, mink, and rats. A variety of species may coexist in one colony by occupying different parts of the habitat. Guillemots squeeze into crevices at the bases of cliffs, while kittiwakes nest above on small rock ledges;

Long, blonde plumes on the heads of breeding tufted puffins (Fratercula cirrhata) *give this member of the auk family its name. Nesting in burrows on isolated islands from California to Alaska, the tufted puffin's heavy beak is used for catching fish and crushing molluscs and sea urchins. Researchers at puffin colonies have been hit on the head by puffins returning to their burrows. This clownish-looking bird winters far offshore: it has been seen eight hundred kilometres from land.*

common murres use higher, larger ledges; puffins dig burrows; and storm-petrels seek the seclusion of forests.

Anyone who visits a seabird colony is invariably impressed by the excessive noise and fragrant fowl air. As fish-eaters, seabirds whitewash the rocks with guano that is rich in nitrogen and phosphate. The droppings, sold as fertilizer in some countries, are sometimes used to glue nests to rocks.

Many seabirds, particularly tubenoses—petrels, albatrosses, shearwaters, and fulmars—are able to sniff through the fishy fetor of the colony and find their own nests by scent. They gather food offshore for days, then locate their nestlings through oversized nostrils. These birds, along with the alcids—auks, murres, puffins, guillemots—also have enlarged nasal glands that adjust the salt content in the blood, allowing them to drink seawater during prolonged offshore forays.

The alcids and some tubenoses catch their prey by "flying" underwater with short, muscular wings, which they use for propulsion and steering. After a hunting expedition they may stand in an afternoon breeze with wings outstretched as if worshipping the warmth of the sun. For some birds the spread-wing posture is, in fact, a method of drying the wings while thermoregulating the body. It also may help to get parasites moving, making them easier to remove.

Not all seabirds nest on the coast. Most gull species nest on inland waterways and many shorebirds prefer the barren Arctic tundra. Some shorebirds—spotted sandpipers, sanderlings, red phalaropes, and others—are polyandrous: females mate with myriad males, lay the eggs, and leave the males to perform all the nesting duties. Appearances are reversed among those that practise polyandry: females are generally larger and more colourful than their mates.

Arctic-nesting shorebirds and seabirds are among the world's great travellers, sometimes visiting four continents a year. Some Arctic terns migrate from Canada across the Atlantic to Europe, then down African shores to wintering grounds in

Frequently confused with cormorants, the anhinga (Anhinga anhinga) *is Canada's only member of the family Anhingidae. Close examination shows its body is sleeker and tail and neck are longer than those of cormorants. Its straight, pointed bill is similar to that of a heron. The anhinga is a native of the tropics and subtropics;* The Birds of Canada, *published by the National Museums of Canada, lists its status here as "accidental in Ontario," based on a record from 1904.*

South Africa or Antarctica, a return voyage of 35,000 kilometres. Sanderlings may fly 24,000 kilometres from northern Canada, down the east coast and across the Caribbean to Chile or Peru. They return up the west side of North America to the Arctic.

Sanderlings, ruddy turnstones, dunlins, stilts, red knots, western and semipalmated sandpipers, and other shorebirds arrive in great flocks to rest at key wetlands along the migration routes. Canada's Bay of Fundy and Fraser delta are of particular importance. They are a crucial link in the chain of staging areas spaced along their migration routes: the loss of one link could be disastrous.

Canada's only penguin—the great auk—became extinct during the last century. Mariners, who used chicks for fish bait, exploited the great auk for its feathers, eggs, and meat. The last of the species was reported off Newfoundland in 1852. Today, seashore birds on Canada's official endangered-species list include the Eskimo curlew and the piping plover.

The belted kingfisher (Ceryle alcyon), *found almost everywhere in Canada below 60 degrees north latitude, is the country's only member of the kingfisher family. A nonmigratory bird, it takes its prey by diving headlong from branches hanging over water. Like an osprey, it can hover before diving. The kingfisher nests in burrows in banks and road cuts where it lays eggs amid an assortment of regurgitated fish bones.*

American white pelicans (Pelecanus erythrorhynchos) *often work together in flocks to herd fish into shallow water. Their bill pouches can hold about eleven litres of water—three times as much as their stomachs. The purpose of the hard, vertical plate on top of the bill is unknown. White pelicans occur mainly in the Prairie Provinces, where they feed on fish, frogs, and other aquatic life. After a worrisome decade, Canada's white pelican was relieved of its official "threatened" status in 1986.*

The pigeon-sized American black oystercatcher (Haematopus bachmani) *is a Pacific shorebird distinguished by its pure black body and bright crimson bill and ring around its yellow eye. The end of its bill is vertically flat, a tool it inserts into oysters, mussels, and other bivalves to sever the adductor muscle, forcing the shells apart.*

Above: The red knot (Calidris canutus) *breeds on the vegetated, soggy tundra of Ellesmere and Prince Patrick islands and other far-north habitats. It probes mud flats and beaches with its pointed bill for snails, periwinkles, worms, beetles, crane flies, fish, seeds, and the eggs of small crabs. At summer's end many red knots head for southern Chile's Straits of Magellan, fifteen thousand kilometres from their Arctic nesting grounds.*

Opposite: The least sandpiper (Calidris minutilla), *between twelve and seventeen centimetres long, is the smallest of Canada's sandpipers. Despite its diminutive size, it is a tireless migrator, travelling each year from its subarctic nesting grounds to winter in the West Indies, Peru, and Brazil.*

Common murres (Uria aalge) *and black-legged or Atlantic kittiwakes* (Rissa tridactyla) *share a colony at Newfoundland's Witless Bay Bird Sanctuary. About a hundred thousand murres nest on Green Island, one of three islands in the 140-hectare refuge, a favourite summer touring area for charter boats.*

Above: The Arctic tern (Sterna paradisaea) *probably migrates farther than any animal on Earth. Some travel from Arctic Canada, across the Atlantic to Europe, and down the African coast to Antarctica, a return migration of 35,000 kilometres. Arctic terns banded in Canada have been seen in France and South Africa. One banded as a nestling in a Norwegian colony died in the same place twenty-seven years later.*

Overleaf: Western sandpipers (Calidris mauri) *gather in huge flocks as they migrate between northern nesting grounds and wintering areas in the southern United States, West Indies, Peru, and Venezuela. One of Canada's three smallest sandpipers, they are seen in many parts of the country, mainly the west.*

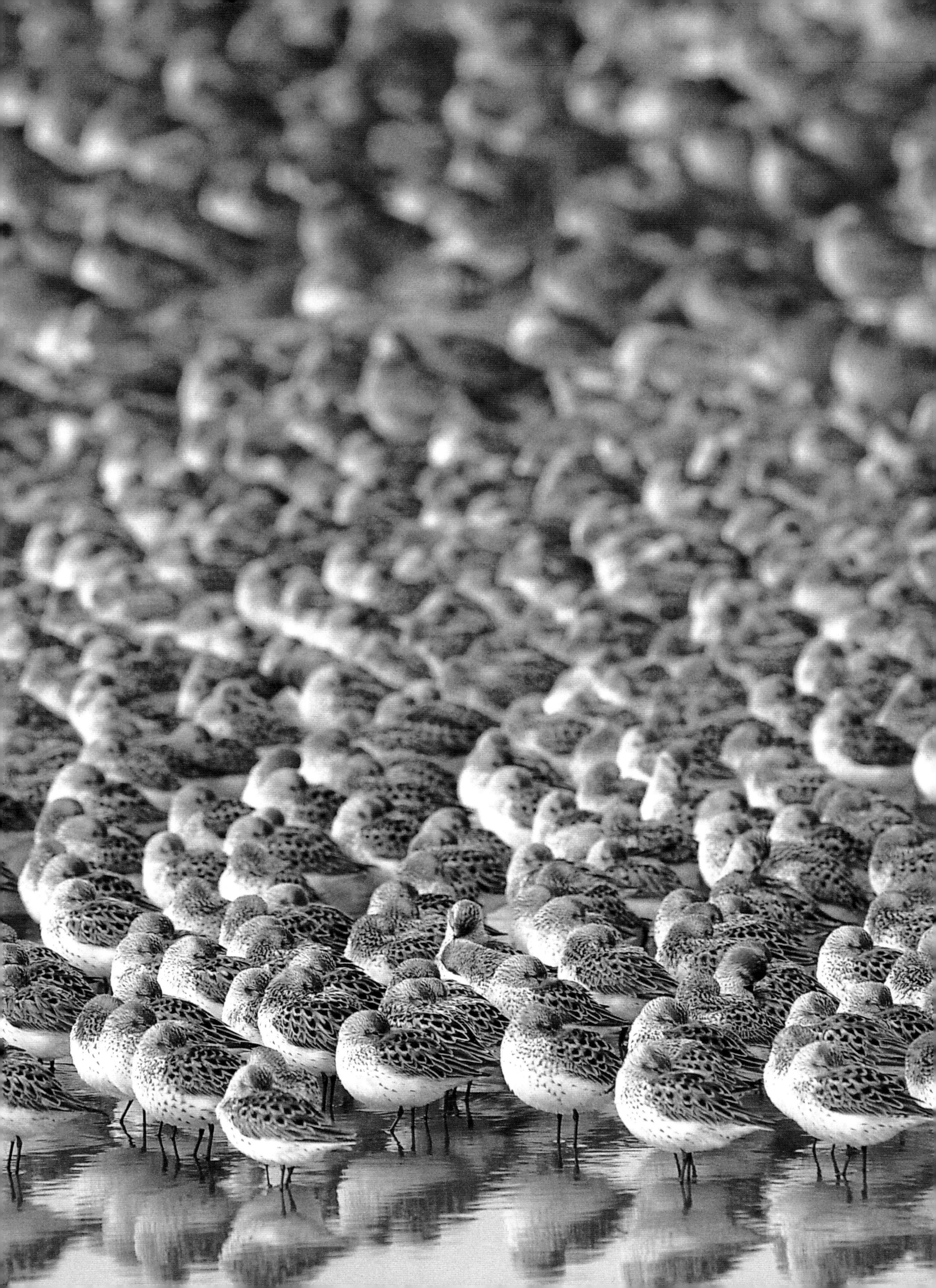

The pomarine jaeger (Stercorarius pomarinus) *breeds on wet mossy tundra of the Arctic. It is one of the main predators of lemmings, and its breeding success is tied closely to fluctuations in lemming populations. Though pomarine jaegers may nest inland, they travel far offshore when migrating to wintering grounds in the West Indies, South America, and the Galapagos Islands.*

The sandhill crane (Grus canadensis), *with its distinctive rusty forehead, is found from British Columbia to eastern Ontario and north to Devon Island, seventeen hundred kilometres north of the Arctic Circle. The elaborate courtship dances of sandhill cranes—head bobbing, leaping, running with extended wings—are imitated in the dances of North American native Indians.*

Above: The smallest of Canada's three jaeger species, the long-tailed jaeger (Stercorarius longicaudus) *is an Arctic breeder. Relatives of gulls, jaegers are aerial pirates, harassing other birds until they drop their prey. The long-tailed jaeger feeds mainly on rodents, birds, insects, fish, and bird eggs.*

Opposite: The American bittern (Botaurus lentiginosus) *is designed to fade into its background. Striped and coloured like marsh reeds, it extends its head upwards and stands motionless when threatened. Its call, like the sound of an old-fashioned water pump or a stake being driven into the ground, has led to the nicknames "thunder pump" or "stake driver."*

Above: A great blue heron (Ardea herodias) *hunts in the shallows. These long-legged waders eat fish, frogs, rodents, and other small animals in fields, marshes, and saltwater lagoons across most of southern Canada. Standing well over a metre tall, they are Canada's largest herons, sometimes living more than twenty years.*

Previous page: These herring gull (Larus argentatus) *chicks will be old enough to wander from their nest in about five days. Adult herring gulls that nest on the ground reject stray chicks that wander into the wrong nest. Herring gulls nesting on cliffs, where young fledge at an older age, accept foreign chicks that are one or two weeks old.*

Double-crested cormorants (Phalacrocorax auritus) *distinguished by their orange throat patches, are the only cormorants likely to be found in Canada's interior as well as on the coast. They share nesting colonies with gulls and other seabirds. Cormorants prefer to nest and roost on the windward side of islets, where they can take flight into the wind.*

Above: The future is precarious for Canada's endangered piping plover (Charadrius melodus)*, whose numbers have been severely depleted by hunting and habitat loss from human settlement. It is found now only in pockets of the southern Prairies, the Niagara Peninsula, Nova Scotia, Prince Edward Island, and a few other scattered locations.*

Opposite: At more than 140 centimetres, the whooping crane (Grus americana) *is Canada's tallest bird. It is also one of North America's most endangered, with a total population probably lower than two hundred. After nesting on the Prairies it migrates to Texas or New Mexico. Through a captive breeding program, using sandhill cranes as foster parents, Canada and the United States have been working together since the 1940s to rescue the whooping crane from extinction.*

The American avocet (Recurvirostra americana) *in Canada is a resident of the southern Prairies, where it's found on the shores and in the shallows of thinly vegetated sloughs, ponds, and alkaline lakes. It catches crustaceans and aquatic insects by sweeping its long, upturned bill back and forth beneath the water, feeling for its prey. Feeding avocets are often followed by Wilson's phalaropes, which grab morsels not eaten by the avocets.*

Though the killdeer (Charadrius vociderus) *is a shorebird, it prefers open areas some distance from water, where its black and white stripes are camouflaged by stones and gravel. It feigns a broken wing to distract predators from its nest. This bird, one of the earliest spring migrants to return to southern Canada, is named for its high-pitched "kill-dee" call.*

Above: The most widely distributed of Canada's twenty-two gull species, the herring gull (Larus argentatus) *is found all over the country except in the high Arctic. This is the ubiquitous "seagull," a large, drably coloured scavenger and predator of other birds and eggs. Its range is expanding with an increase in human garbage.*

Opposite: Slightly smaller than a great blue heron, the snow-white great egret (Casmerodius albus) *breeds in Canada only at a few specific sites in southern Saskatchewan, Manitoba, Ontario, and Quebec. Larger than a snowy egret, it occasionally visits the Atlantic provinces, Alberta, and the southwest coast of British Columbia.*

The black-legged kittiwake (Rissa tridactyla) *nests in crowded colonies on steep cliffs. In Alaska's Pribilof Islands, literally millions nest on a sheer three-metre-high cliff that stretches along eight kilometres of coastline. These colonies are often so tight that only one parent can squeeze onto the nest at a time. The kittiwake not only calls its own name, but recognizes its mate and offspring by sound.*

Though separated by nearly five thousand kilometres, the Atlantic puffin (Fratercula arctica) *is a close cousin of the Pacific's horned and tufted puffins. It is capable of carrying up to thirty small fish, which are snared with small pinchers at the tip of the multi-coloured bill, then pushed back by the tongue. The fish are held in place by serrations on the upper mandible. Eating its body weight each day on the nest, a puffin chick might consume two thousand fish before fledging.*

WATERFOWL

The Canadian Prairies are the breeding grounds for nearly three-quarters of North America's waterfowl. More than 750,000 square kilometres across Alberta, Saskatchewan, and Manitoba are a myriad of potholes, sloughs, and sluggish streams, rich with insects, molluscs, crustaceans, amphibians, and aquatic vegetation. This is prime habitat for ducks, geese, and other waterbirds.

It is on these vast wetlands that Ducks Unlimited, a private U.S. conservation organization that came to Canada in 1938, has done most of its work. Since completing its first enhancement project at Manitoba's Big Grass Lake, Ducks Unlimited has spent nearly $300 million on three thousand wetland projects involving 1.5 million hectares, primarily on the Prairies.

The marshes of Manitoba are known internationally: among the nation's most productive waterfowl breeding areas are

At only thirty-six centimetres long, the green-winged teal (Anas crecca) *is the smallest Canadian dabbling duck. Despite its size, it was common for nineteenth-century hunters to shoot half a dozen a day as the first migrants arrived. Young green-winged teals grow faster than any other North American duck.*

Minnedosa Potholes and Delta Marsh, major destinations for mating mallards, ring-necked ducks, scaups, pintails, green and blue-winged teals, shovellers, canvas-backs, redheads, and more.

In neighbouring Saskatchewan, the first bird sanctuary on the continent was established in 1887 at Last Mountain Lake. A century later, nearby Quill Lakes became the first preservation project under the North American Water-fowl Management Plan, a 1986 Canada-U.S. agreement to protect and enhance 1.5 million hectares of wetlands by the turn of the next century.

East of the Prairies, the marshes and waterways of the Canadian Shield stretch across much of Ontario and Quebec. The upper St. Lawrence River and the land surrounding Lakes Ontario, Huron, and St. Clair are dominated by wetlands. One of the most prominent is Minesing Swamp, eighty-one square kilometres near the north shore of Lake Ontario.

It is at these wetlands that birders are likely to see the courting rituals of ducks. Mallards, the most pervasive of water-birds, float about looking into each other's eyes, bobbing their heads before copulating. A female encourages potential partners to fight for its affections by swimming with its head just above the surface, like a periscope, barking short quacks while waving its beak. A drake raises its chest and vibrates its head and tail while swimming around the chosen hen.

Males of many duck species abandon their partners after breeding and fade into eclipse plumage: their colourful con-nubial attire moults, rendering the birds dull and flightless. Some remain in this vulnerable state until the following spring.

Like seabirds, a variety of waterfowl shares a wetland by using different areas. Dabbling ducks such as mallards, pin-tails, and teals forage on vegetation along the edges of ponds. Divers like scaups, redheads, or canvasbacks prefer deeper water at the centre of a pool, where they swim for snails and aquatic insects. Divers and dabblers can also be distin-guished from one another by their flight:

Canvasbacks (Aythya valisineria) *are among those unfortunate ducks that are susceptible to lead poisoning. Like mallards, pintails, redheads, and Canada geese, they often forage in fields for seeds that resemble lead shot. Shortly after hunting season they can be seen stumbling across contaminated wetlands, gradually losing half their body weight before dying.*

larger-winged dabblers need less space to take off and land and can fly from a pond almost vertically; the shorter wings of divers require longer runways where they can flap their webbed feet across the surface to become airborne.

The oversized feet of waterbirds are used to propel themselves while swimming both above and below the surface. Perhaps the most proficient diving bird is the loon, capable of catching fish at depths to 180 metres. Loons, like old-squaw, grebes, harlequin ducks, mergansers, goldeneyes, and buffleheads, frequent both fresh water and the sea.

One of the pleasures of living in Canada is watching great flocks of geese flying overhead in orderly V-formations. Whether this idiosyncracy is aerodynamically helpful, or provides better vision to avoid collisions, remains a long-standing enigma in the birding world.

The largest North American waterfowl, with a wingspan of two metres, is the twelve-kilogram trumpeter swan. Like geese and smaller waterfowl, swans that forage in fields and marshes may needlessly face the anguish of lead poisoning. After annual hunting seasons the birds ingest shotgun pellets while feeding. Lead flows through the bloodstream, contaminating the liver, kidneys, brain, heart, and other vital organs. The hapless birds linger in pain for two or three weeks before succumbing to the poison. Their carcasses are potentially lethal to bald eagles, turkey vultures, and other scavengers farther up the food chain. Lead shot is banned in some parts of Canada, but the nation is slow to follow the example in the United States, where only steel shot is allowed for waterfowl hunting.

Hunters and human encroachment caused the demise of Canada's Labrador duck, a seagoing relative of eider ducks that enjoyed Atlantic sandbars. Pollution from residential development, collection of eggs, and snaring them with baited hooks contributed to the extinction of the Labrador duck by the late 1800s. Today harlequin ducks in the eastern provinces are Canada's only endangered waterfowl.

The American black duck (Anas rubripes) *is the most common breeding duck in southeastern Canada. The east's answer to the ubiquitous mallard, it is found through all of Ontario and the Maritimes and most of Quebec. Where mallard and black-duck ranges overlap, the two species hybridize extensively: both are extremely similar in size and appearance. Though plentiful in Canada, there is concern over populations in the United States where they have been reduced by hunting, acid rain, habitat losses, and aerial spraying of pesticides.*

Above: Young Canada goslings (Branta canadensis) *will migrate south with their parents and return as a family to the nesting grounds the following spring. In the mating areas the young of the previous year will form flocks with other non-breeding birds. Most Canada geese breed in their third year, occasionally in their second.*

Previous page: The most pervasive of Canadian waterfowl, Canada geese (Branta canadensis) *nest nearly everywhere except Nova Scotia and the high Arctic. Their wavy, V-shaped flocks and incessant honking are welcomed across the nation as they migrate to and from their nesting grounds. This mating pair, together for life, may produce half a dozen goslings.*

A gadwall (Anas strepera) *breaks from foraging the shallows to stretch its wings. Slightly smaller than a mallard, the gadwall is a dabbler that survives almost entirely on aquatic vegetation. It is mainly a southern breeder in Canada, found in British Columbia and the Prairies, the Niagara Peninsula, and Prince Edward Island. Though widespread in North America, it is not numerous.*

Above: A mallard drake (Anas platyrhychos) *comes in for a landing. Dabbling ducks, with larger wings than divers, can take off and land almost vertically. The most widespread duck in the northern hemisphere, the mallard is highly adaptable, feeding in ponds, marshes, lakes, streams, fields, and saltwater.*

Opposite: Brood parasitism—putting your eggs in someone else's basket—is common among redheads (Aythya americana). *In one Alberta study of nearly seven hundred nests, one-fifth were parasitized by redheads. The reasons for this odd behaviour are uncertain: perhaps the birds are driven by a shortage of nesting sites; or it could be that young ducks, more independent than other species, rely less on their mothers for food and protection, allowing large numbers to be raised by a single parent.*

The harlequin duck (Histrionicus histrionicus) *is the wood duck's seagoing rival. Its colourful plumage is a sight that brings birders, artists, and photographers to rocky seashores in winter. In spring the harlequin moves inland to nest on swift-flowing streams. The brilliantly coloured males lose their artistic appeal during summer moult.*

Above: Mainly an inhabitant of the western provinces, the crow-sized ruddy duck (Oxyura jamaicensis) *is a diver that feeds on insect larvae and aquatic snails and vegetation. Females often lay their eggs in the nests of redheads, canvasbacks, grebes, and rails. Adult males can be identified by their blue bills.*

Overleaf: After wintering as far south as Mexico, Canada's snow geese (Anser caerulescens) *return to the high Arctic to nest. They gather in colonies, sometimes with brant, and build down-padded nests of moss and other tundra vegetation. Eastern Canada's snow-goose populations are estimated at 2.5 million and growing by 130,000 a year. Migrations are spectacular, with large flocks flying over James and Hudson bays and more than 100,000 stopping near Quebec City at Cap Tourmente National Wildlife Area. The migration of Canada's only wintering snow geese, about 40,000, peaks at the Fraser estuary in November.*

Above: The call of the loon, as singular as the howl of a wolf, is a symbol of Canada's northern wilderness. The common loon (Gavia immer) *is the most widespread of the country's five loon species, found almost everywhere except the high Arctic. The common loon, nearly a metre long, is a diving bird, catching fish at depths of 180 metres.*

Opposite: Found in large numbers across Canada, the American wigeon (Anas americana) *is particularly abundant on the west coast, where as many as 900,000 a year migrate on the Pacific flyway and 60,000 winter on the Fraser delta. These dabblers keep the company of American coots, which dive deeper. When food is scarce in the shallows, wigeons knock food from the bills of coots and steal it.*

Once known as the whistling swan for its peculiar three-note call, the tundra swan (Cygnus columbianus) *breeds in the south and mid-Arctic. A flock in flight sounds like Canada geese but looks like a diminutive version of trumpeter swans, which are about 25 percent larger. When migrating north in severe weather, ice often collects on the wings of tundra swans.*

Above: The dark band encircling the neck of an adult male ring-necked duck (Aythya collaris) *is barely discernible. Like the mallard, this species is widely distributed in Canada. And like other ducks, the ring-necked is susceptible to poisoning from lead shot left in fields and wetlands after hunting seasons.*

Overleaf: With a one-metre wingspan, the common merganser (Mergus merganser) *is the largest inland duck on the continent and the biggest of Canada's three merganser species. An expert diver, the merganser eats fish, catching its prey with a long, serrated bill. Hollow trees, rock piles, and holes in banks make ideal nest sites for common mergansers.*

Above: Heavy-billed surf scoters (Malanitta perspicillata) *swim in the near-shore waters off the coast of B.C.'s Queen Charlotte Islands. These duck-sized divers breed across northern Canada and winter along the east and west coasts of North America. In winter and spring, during Pacific herring spawns, they gather in flocks of up to 300,000.*

Opposite: Measuring less than forty centimetres long, the bufflehead (Bucephala albeola) *is North America's smallest diving duck. It nests in hollow trees across most of Canada west of Quebec and is most abundant in the woodlands north of the Prairies. Its unusual name is a reference to its oversized head, like that of a buffalo.*

The inharmonious discordance of the blue-winged teal's call is probably the basis of its Latin name, Anas discors. *Flying swiftly in compact flocks, these birds make several passes over a landing site to ensure there is no danger before setting down. When surface-feeding in shallow ponds they are less suspicious, almost tame, and can be approached easily by curious birders.*

The largest North American waterfowl, the trumpeter swan (Cygnus buccinator) *weighs twelve kilograms and spreads its wings to more than two metres. Its snow-white body, from the tip of its tail to its black beak, is nearly as long as its wingspan. This bird's deep, projecting honk is deafening to someone nearby. Trumpeter swans in Canada were severely depleted in the early part of this century. In the United States, where they were slaughtered for food and feathers, their numbers fell to a mere sixty-six birds by 1933. Through stringent protection and captive breeding projects, their numbers now are climbing steadily, but the trumpeter swan remains officially "vulnerable" in Canada.*

Above: The brant (Branta bernicla) *is a small sea goose that breeds in Canada's high Arctic. Sought after by hunters and birders, the arrival of brant at points along the migration route is invariably cause for excitement. Their incessant chatter while feeding on eelgrass along shallow seashores has earned brant the nickname "talking geese." These endearing geese are seriously threatened by loss of feeding habitat. British Columbia has given some brant-feeding areas special protection and banned brant hunting in some parts of the province.*

Previous page: The Barrow's goldeneye (Bucephala islandica) *is found in Canada mainly west of the Rocky Mountains, where it gathers in huge flocks in spring while waiting for ice to melt from smaller lakes and ponds. It nests in tree cavities and migrates comparatively short distances from nesting areas. This small diver is named in honour of Sir John Barrow, secretary to the British Admiralty, who visited North America in the late eighteenth century.*

The king eider (Somateria spectabilis) *is a northern species that breeds close to the sea on shallow ponds of the Arctic tundra. It winters at the northern extremity of ice-free water, mainly on the ocean. Among waterfowl it is the second-deepest diver, next to the oldsquaw, and is capable of reaching depths of sixty metres.*

Birds of Prey

Birds of prey are the carnivores of the avian world. All of Canada's thirty-nine raptorial species—hawks, vultures, eagles, kites, harriers, ospreys, and owls—are superbly equipped to practise their predatory skills. Highly evolved vision, in some cases eight times as acute as that of a human, allows them to ambush their prey with inimitable speed and precision. They grab and dispatch their victims on the wing with powerful, curved talons.

Some raptorial birds have evolved to hunt specific types of animals. Ospreys, nicknamed fish hawks, have spines on the soles of their feet to grip slippery fish. Streamlined peregrine falcons, Canada's fastest birds, stoop toward their quarry at speeds exceeding three hundred kilometres an hour, striking in aerial bursts of shrieks and feathers. The sensitive ears of night-hunting owls can hear a shrew or mouse crawling in a grassy field. An owl's feathers are designed to

With a two-metre wingspan, a bald eagle (Haliaeetus leucocephalus) *soars over its territory. These stately birds, adopted in 1782 as the United States national emblem, are invariably associated with large bodies of water. In coastal areas they build lofty stick nests, some measuring two or three metres across and weighing a tonne, within a hundred metres of shore. Fish, which they catch or steal from other birds, are a major part of an eagle's diet.*

quell the sound of air flowing over a smooth surface. Owls become intimately familiar with the habits of creatures in their territories.

Unlike other birds, which have stereoscopic sight, raptors see with binocular vision—both eyes facing forward. This allows them to judge distances accurately, but impairs their all-round vision. Many raptors have evolved to rotate their heads in almost a complete circle. Some owls can turn their heads in a 270-degree arc. Raptors' eyes are extraordinarily large in relation to their bodies: proportionally, human eyes would have to weigh nearly two kilograms each to match those of some raptorial birds. A raptor's retina may receive twice as much light as a human eye, bringing higher resolution in darker situations.

Some birds of prey gather food by kleptoparasitism, or piracy. Turkey vultures, which rid the environment of rotting carcasses, are known to force heron nestlings to regurgitate their most recent meals. Bald eagles may work in pairs: one bird chases an osprey, forcing it to drop its fresh meal, while the other closes in and catches the falling fish.

Owls and some larger birds of prey swallow their meals whole, digesting all but the bones and feathers. They regurgitate the undigested parts as hard, fuzzy pellets. Pellets found below nests and raptor roosts help ornithologists determine what the birds are eating. In the Canadian north, owls are particularly fond of lemmings, Arctic ground squirrels, and snowshoe hares.

Survival of raptorial birds is closely tied to migratory and biological cycles of prey. About twenty or thirty thousand bald eagles that winter on Canada's ice-free Pacific coast are lured from colder inland areas by spawning salmon, herring, and wintering waterfowl. Goshawks, rough-legged hawks, and great horned and short-eared owls move south from their Arctic domains when there's a crash in populations of rodents. Snowy owls rely heavily on lemmings, which flucutate greatly in numbers, reaching population peaks about every four years, then crashing. As a result large numbers

Stooping toward its airborne prey at speeds of three hundred kilometres an hour, the peregrine falcon (Falco peregrinus) *is North America's fastest bird. This majestic raptor has suffered the effects of DDT and other pesticides more than any other North American bird of prey. In Canada the subspecies* anatum, *which includes most of the southern breeders, is officially endangered. The subspecies* tundrius, *of the Arctic, is "threatened," and the subspecies* pealei, *of the Queen Charlottes and Moore Island in British Columbia, is listed as "vulnerable."*

of snowy owls seek alternate prey in southern Canada approximately every four years.

Unable to store fresh flesh, raptorial birds must hunt year-round. Some owls, however, wait for freezing winter weather to stockpile bugs and small mammals. To thaw their dinner, they simply sit on the meal and incubate it like an egg.

Partners for life, many raptors raise their offspring in lofty stick nests, which they repair and reuse year after year. Parasites are reduced by adding vegetation such as cedar bark or green leaves, which have pesticidal properties. Heavily parasitized nests may be abandoned.

In ancient art and mythology, birds of prey portray power, evil, or divinity. Owls in early times were associated with wizardry and witchcraft. In Greek and Roman tradition the eagle is the revered bird of Zeus; it embellishes the legends and totems of Canadian native Indians as the mighty thunderbird. In 1782 the bald eagle was adopted as the U.S. national emblem. And today's Middle Eastern aristocracy is still fascinated by falconry, a sport dating back to 1200 BC.

It wasn't until this century that the stately image of birds of prey faded. As Europeans settled across North America, raptorial birds were shot on sight as vermin—sheep and poultry killers to be annihilated along with any other predator. Most governments paid bounties for dead raptors.

The greatest destruction of raptorial birds occurred in the mid-1900s and was inadvertent: dichlorodiphenyl trichloroethane, a potent synthetic pesticide better known as DDT, was used heavily across the continent after the Second World War. As tonnes of DDT were spread across North American farms, raptor populations, particularly eagles and peregrine falcons, began to crash. They were almost extirpated from the east. They fared better west of Saskatchewan but declines were serious. DDT, which reduces the calcium content in eggs, causing them to break, was banned in the United States at the end of 1972. In Canada it was severely restricted in 1970, prohibited in 1990.

At about twenty centimetres long, the northern saw-whet (Aegolius acadicus) *is eastern Canada's smallest owl: only the pygmy-owl of B.C. and western Alberta is smaller. Named for one of its calls that resembles the filing of a saw, this minuscule raptor is drawn to people by insatiable curiosity. Many are caught by hand as they roost during the day.*

Today, with the help of transplants, artificial nests, pollution controls, lead-shot bans, and captive breeding and release programs, raptorial birds are repopulating traditional habitats. Only one subspecies of peregrine falcon remains on Canada's endangered-species list.

One of the largest and strongest of North American owls, the great horned (Bubo virginianus) *was the "eagle owl" of Pliny. It seems that few animals are safe from its formidable hunting skills: shrews, rabbits, woodchucks, skunks, weasels, geese, swans, herons, domestic turkeys, pet cats, even porcupines. This bird is known to descend chimneys in pursuit of swifts. Found across the country, its movements in the north are closely tied to the biological cycles of snowshoe hares.*

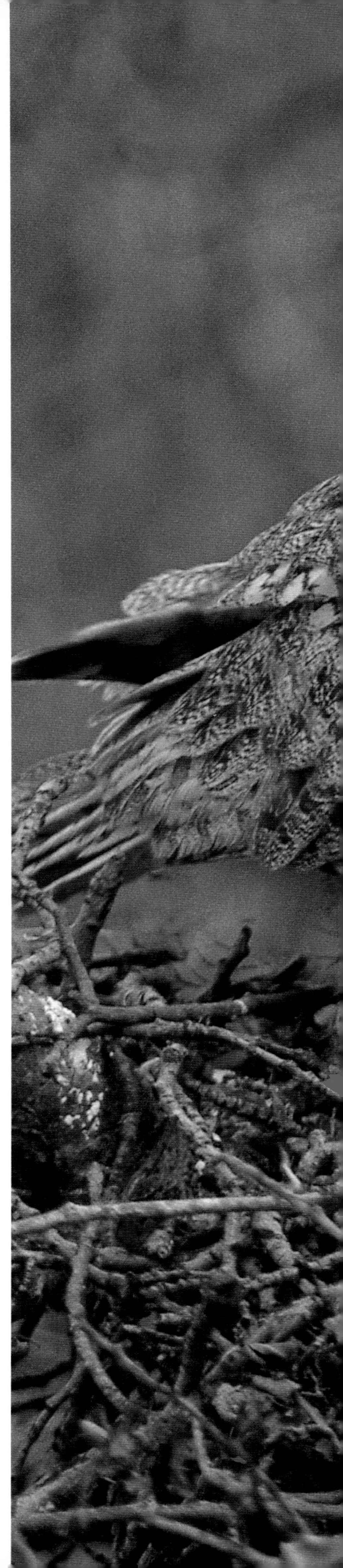

The long-legged burrowing owl (Athene cunicularia), *occurring in Canada only on the grasslands of southern British Columbia and the Prairies, is officially "threatened." It nests in burrows abandoned by mammals and imitates the rattle of a rattlesnake when its young are jeopardized. Efforts to save this vanishing species from extirpation by introducing unfledged young from the United States have met with dubious success.*

Drifting as high as a kilometre in the air on afternoon thermal currents, turkey vultures (Cathartes aura) *survey areas of three square kilometres or more. They stay aloft all day, reaching speeds of sixty kilometres an hour with near-imperceptible movement of their wings, which spread to nearly two metres. They descend only when they spot a dead deer, seal, sea lion, skunk, rabbit, or other carcass. These big raptors are found in southern Canada from British Columbia to southwest Ontario and on the Niagara Peninsula.*

Above: The distinctive white head and tail of the bald eagle (Haliaeetus leucocephalus) *is acquired as the birds mature at four or five years old. An adult bird may measure almost a metre from tail to head and weigh between 2.5 and 6.3 kilograms. As with most raptorial birds, females are about one-third larger than their lifelong partners.*

Previous page: Though the rough-legged hawk (Buteo lagopus) *nests in Canada's Arctic and subarctic regions, it migrates south to more populated areas in winter. Its Latin name,* lagopus, *or "hare's foot," is derived from the furlike feathers that cover its legs to its toes.*

Shorter than twenty-three centimetres, the American kestrel (Falco sparverius) *is the smallest and most common North American falcon. Its nickname "sparrow hawk" is a misnomer: it is not a hawk and it eats sparrows only occasionally. The kestrel's penchant for mice and grasshoppers makes it a friend of the farmer.*

Above: The osprey (Pandion haliaetus) *is the epitome of form and precision. Surviving exclusively on fish, the "fish hawk" soars over the water: when it spots a fish it spreads its tail feathers and hovers. Then, from a height of fifteen to thirty metres, it plunges feet-first into the water, often disappearing completely. Other times it zeroes in on a fish and snatches it from the surface in one quick skim. This proficient hunter is successful more than half the time.*

Opposite: Adult red-tailed hawks (Buteo jamaicensus) *are easily recognized by their rufous tails: the tips of the feathers are light-coloured with a dark band. Birders also identify these handsome raptors by their harsh, prolonged scream. These fearless birds are known to attack humans who approach their nests, raking intruders with their talons. Young of the year, however, may be quite tame, making them easy targets for trigger-happy raptor haters.*

Above: The western screech-owl (Otus kennicottii), *with its large yellow eyes and short ear tufts, flies silently at night on broad, rounded wings and sits in seclusion by day, usually in a crevice or tree cavity. Its call is not actually a screech but an accelerating series of hollow, evenly pitched whistles, ending in a whinny. These birds readily respond to imitated calls.*

Previous page: An inhabitant of the arid grasslands and hills of southern British Columbia, Alberta, and Saskatchewan, the prairie falcon (Falco mexicanus) *is on the decline in Canada. It stoops to its prey at speeds rivalling that of the peregrine falcon, knocking a hapless bird from the air with its powerful feet or snatching it in its talons. Like kestrels, kites, and red-tailed hawks, the prairie falcon often hovers over its intended meal before attacking.*

Once known as marsh hawks, northern harriers (Circus cyaneus) *frequent the plains, sloughs, wet meadows, and marshes of southern Canada. They look somewhat like a cross between an owl and a hawk, with an owllike facial ruff that reflects sound. By flying within two metres of the ground, harriers can detect voles, snakes, frogs, insects, birds, and other prey by sound.*

Above: Flying fast and low, hugging the contours of the land, the gyrfalcon (Falco rusticolus) *takes its prey—ptarmigan, grouse, seabirds, waterfowl, shorebirds, and other animals of the Arctic—by surprise. The largest of all falcons, this stately raptor is the precious prize of Middle East royalty, who began practising the sport of falconry more than three thousand years ago. Wild Canadian falcons are illegally shipped to overseas falconers, but the extent of the trade is not known.*

Opposite: A bald eagle (Haliaeetus leucocdephalus) *guards its eaglets in the nest. This distinctive bird of prey is found in the Yukon and across southern Canada except on the southern Prairies. It is particularly abundant, and resourceful, on the Pacific coast, where twenty or thirty thousand winter near the ice-free waters. They feed on spawning herring and salmon and a multitude of waterfowl. Eagles in the Queen Charlotte Islands have been known to stalk the seashore, pulling back seaweed at low tides to tear abalone from rocks.*

Above: Unlike some smaller owls, the large and powerful snowy owl (Nyctea scandiaca) *is leery of humans, rarely allowing them to approach nearer than about ninety metres. An Arctic nester, the snowy owl encounters people only when northern lemming populations crash, about every four years, forcing the owls to hunt prey in more populated southern regions.*

Previous page: Aboriginals, who believed the boreal owl (Aegolius funereus), *found across most of Canada, could be captured easily because of its poor eyesight called it "the blind one." This small owl, in fact, can be approached because it is unafraid of humans. Its Latin name,* funereus, *refers to a bird regarded as an ill omen, a reputation earned by its melancholy night cry, a single note repeated every minute or two.*

The pretty plumage of the red-shouldered hawk (Butea lineatus) *is becoming a rare sight in Canada. Listed as "vulnerable" due to declining numbers, it is found now only in small southeast sections of Ontario, Quebec, and New Brunswick. It prefers valley bottoms and swamps, where it hunts rodents, snakes, lizards, and insects in meadows and on the edges of forests. It is susceptible to pesticide poisoning but is threatened more by habitat loss.*

Above: The golden eagle (Aquila chrysaetos) *occurs across most of Canada, except the high Arctic, and is particularly fond of open areas in mountains and hills. Though widespread, it is not plentiful, especially in the east, where many of its traditional breeding sites are abandoned. Like many birds at the top of the food chain, the golden eagle suffered severely from DDT poisoning earlier in this century.*

Opposite: The common barn-owl (Tyto alba), *with its small eyes and heart-shaped face, is the only member of the family Tytonidae. Though it differs in appearance from other owls, its habits are much the same. Often a city dweller, the barn-owl is not widespread in Canada, breeding only in southwestern B.C. and southeastern Ontario. It is a formidable hunter, capable of hearing the movements of rodents in fields.*

Above: The Swainson's hawk (Buteo swainsoni) *is the most common large hawk of the Prairies, where it surveys open territories from fence posts. Rodents are its prime prey and the Swainson's hawk is a valued controller of mice, gophers, and grasshoppers. After breeding on the Canadian Prairies it usually migrates more than eleven thousand kilometres to Argentina. This raptor was named in 1838 for William Swainson, a brilliant English ornithologist.*

Previous page: The crow-sized short-eared owl (Asio flammeus) *occurs virtually everywhere in Canada except the high Arctic. Its conspicuous habit of gathering in groups at dusk to hunt mice in fields and marshes makes it one of the most frequently seen owls. It often scouts for prey from fence posts, pouncing on its victims with outstretched wings.*

Like most members of the family Strigidae, the northern hawk-owl (Surnia ulula) *migrates only short distances from its nesting site. The eggs of this species and other owls hatch asynchronously, a characteristic that ornithologists believe evolved as a means of ensuring survival of the fittest. With eggs hatching at different times, the first to hatch and grow are better able to beg for food from parents. The later, smaller hatchlings survive only in years when food is abundant enough for the entire brood.*

Hummingbirds, Woodpeckers, and Upland Birds

Ornithologists often rely on sound, rather than sight, to identify certain species. The call of an owl or cooing of a pigeon, the song of a warbler or throaty croak of a great blue heron are familiar to most birders. But many species can be identified by sounds other than vocalizations.

A hummingbird, so called because of its sound, beats its wings at sixty to eighty times a second, producing a low-pitched buzz like a giant bee. While a birder uses the sound to identify the bird, the hummer uses it in courtship and territorial defence. The rapid-fire tapping of a woodpecker on hollow trees, house gutters, drainpipes, or garbage cans is also used to claim a mate or territory. Mating and territorial ownership

Occurring only in B.C. and southwestern Alberta, the rufous hummingbird (Selasphorus rufus) *forages in various habitats from ocean coasts and islands to forests and high-elevation meadows, feeding on the nectar of flowers. Its extremely efficient kidneys retain body salts while removing unneeded water. About three-quarters of its body weight in fluids is passed each day; humans passing a similar amount would need to expel approximately seventy-five litres a day.*

are the purpose of the muffled drumming sounds that ruffed grouse create by beating their wings upward and forward.

Producing a hummingbird's prolonged hum requires extraordinary energy: this bird not only rivals the shrew as the world's smallest warm-blooded animal, but has the greatest energy output of any bird or mammal. A human would need to consume 130 loaves of bread or 155,000 calories a day to match the energy expended by a hummingbird.

The buzz is best heard as the hummer hovers near a flower or feeder, consuming its body weight in nectar each day. No other bird can match the hovering efficiency of a hummingbird, holding its body stationary at a forty-five-degree angle, moving its wings in a figure-eight pattern. About a third of its diminutive form is breast muscles that drive the wings.

A hummingbird's heart may beat a thousand times a second when buzzing about its daily business. It uses eight times as much energy when active as when resting: it would die of starvation within hours if it stopped feeding. At night, however, a hummer may fall into a deep torpor, allowing its body temperature to drop by half.

Of Canada's five hummingbird species, the ruby-throated is the most widely distributed and the only one found in the east. The rufous hummingbird occurs only in western Alberta and B.C.

Like hummingbird, the name woodpecker is also somewhat related to the noise the bird makes. Each of Canada's fourteen woodpecker species is equipped with a hard, chisellike bill and heavy skull: an industrious insect-eater, it ceaselessly hammers holes in dead or dying trees. The nostrils are protected from flying sawdust by bristlelike feathers.

These short-legged climbers are better adapted than any other species for life on tree trunks. Capped with sharp, curved claws, their feet have two toes in front, one behind, and a fourth turned at a right angle to the trunk. Stiff tail feathers that brace woodpeckers against trunks also propel the birds as they climb.

Wood-boring bugs are extracted with

Yellow-bellied sapsuckers (Sphyrapicus varius) *drink sap from wells they drill in trees and guard from other birds. Young are taught to suck sap as they fledge, making them independent from parents within one or two weeks. In North America they take sap from more than 240 tree species. Sapsuckers also cache nuts and fruit.*

long, wormlike tongues coated with sticky saliva and tipped with barbs. Feeding on beetles, ants, aphids, flies, or caterpillars, woodpeckers are often seen with heads cocked to one side, listening to the movements of their prey. Woodpeckers that eat acorns, pine seeds, nuts, or berries may hoard food and vigorously defend their caches. Lewis's woodpeckers husk acorns before stashing them.

The pileated, with a bright red crest, is the largest Canadian woodpecker. Like a few others, pileated woodpeckers may carry eggs to a new nest site if the original nest is lost or damaged.

While woodpeckers bang holes in trees, galliformes scratch at the ground, unearthing seeds, grains, and bugs. Related to chickens, the upland species—grouse, ptarmigan, pheasants, partridges, quail, and wild turkeys—comprise many of Canada's game birds. Among the most intriguing are sharp-tailed, sage, and other grouse that return each year to traditional sites, known as leks, for elaborate communal courtship displays.

Between twenty and seventy males strut about the lek, tail feathers spread and wings vibrating, inflating air sacs on their breasts. They bob their heads while gobbling and making popping sounds. Male grouse divide the lek into territories, with the dominant male taking the central position and usually enjoying the most active sex life. Most offspring are produced by a minority of hierarchal males. Females select a mate from the lek, breed with it, then nest alone. The lek dances of sage grouse have been well documented on the grasslands of southeast Alberta and southwest Saskatchewan in April and May. Greater prairie chickens held similar courtship meetings in the southern Prairies, but that species now is extirpated from Canada.

In the north the grouse's cousin, the ptarmigan, has evolved to endure harsh Canadian winters. In the snow the pigeon-sized white-tailed ptarmigan, for example, is camouflaged from enemies by pure white plumage. This tundra nester grows long claws and thick tufts of feathers on both sides of its feet to walk

The white-tailed ptarmigan (Lagopus leucurus) *is a western species, living above the tree line in British Columbia and the Yukon. The smallest Canadian ptarmigan, the white-tailed is usually a solitary animal except during breeding. Females, which are left to incubate eggs alone, are extremely protective, often refusing to leave the nest until literally touched by an intruder.*

on snow. The special footgear reduces the distance the foot sinks in the snow by half and increases the bearing surface of the foot by about 400 percent. The white-tailed ptarmigan, like the willow and rock ptarmigan, defends its offspring by distracting intruders while the chicks hide in the bush.

No grouse, ptarmigan, or related upland birds are endangered in Canada, but according to the Committee on the Status of Endangered Wildlife in Canada, the greater prairie chicken now has been extirpated from the country, a victim of human settlement.

The courtship of the male sage grouse (Centrocercus urophasianus) *is perhaps the most spectacular of all North American grouse species. It struts about with tail feathers spread and air sac inflated on its breast, emitting loud popping noises. These rituals are performed on traditional leks, where several grouse gather year after year. Though permanently established in southeast Alberta and southwest Saskatchewan, sage grouse numbers are not high.*

Only residents of southern Vancouver Island and the mainland of south-western B.C. enjoy the soothing call of the California quail (Callipepla californica). *A native of the Pacific coast from northern Baja to Oregon, this plump, black-crested ground-dweller was introduced to Canada around the turn of the century. It forages in open woodlands and shrub, with males standing as sentry while females tend the chicks.*

Perched on its drumming log, a male ruffed grouse (Bonasa umbellus) *attempts to seduce a mate. Known by some people as a partridge, this grouse beats its wings forward and upward, creating a dull drumming that increases in speed before tapering to silence. Ruffed grouse lay low in second-growth woodlands, often startling hikers who almost step on them before the birds take flight.*

Above: Found across southern Canada except in British Columbia, the ruby-throated hummingbird (Archilochus colubris) *is one of five Canadian hummingbird species. It is the most widely distributed hummer in the country and the only one found in the east. In its aerial courtship displays, the male swings like a pendulum before the female, then quickly moves above her and buzzes to each side of her. Face to face, they alternately descend, then drop to the ground and mate.*

Opposite: The northern flicker (Colaptes auratus) *spends more time on the ground than any other woodpecker. It scratches about in search of ants, its favourite food, as well as seeds, acorns, nuts, and grain. Found across most of Canada, except the Arctic, it is a highly adaptable bird, preferring open habitat with nearby tree cover. Yellow-shafted and red-shafted flickers are subspecies of the northern flicker.*

A larger version of the downy woodpecker, the hairy woodpecker (Picoides villosus) *inhabits similar areas. Hairys are not as friendly as downys, avoiding human approaches by dodging behind a tree trunk or fading into the forest. More active and noisier than downys, hairy woodpeckers tend to stay on the same territory for life.*

Like similar species, the sharp-tailed grouse (Tympanuchus phasianellus) *gathers on traditional leks in spring for fancy dances in which the males vie for the affections of females. Starting at sunrise, with heads lowered, wings drooped, tails erect, they furiously stomp their feet while issuing subtle booming sounds through inflated purple air sacs on the sides of their necks. The sharp-tailed is the most common grouse of the Prairies.*

Above: The chicken-sized willow ptarmigan (Lagopus lagopus), *a northern nester, is one of three Canadian ptarmigan species. These Arctic grouse are camouflaged by their surroundings throughout the year as they change colour with the seasons. This species was successfully introduced to Nova Scotia's Scatarie Island in 1968.*

Opposite: The pileated woodpecker (Dryocopus pileatus), *with its unique red crest, is Canada's largest woodpecker. Found in woodlands across southern Canada, this heavy-billed bird requires large territories, which it shares year-round with its mate. The pileated woodpecker readily adapts to reforested areas.*

A resident of most of southern Canada, the downy (Picoides pubescens) *is the smallest woodpecker in the country. Bolder than some of the larger species, it is easily attracted to suet and will roost, but not nest, in artificial boxes. Three-quarters of its diet is insects, and this bird is a major controller of pests in various parts of the country. Some downy woodpecker pairs stay together for several years.*

Photographs are all that remain of Canada's prairie chicken (Tympanuchus cupido), *which now is extirpated from the country. This bird once thrived with the bison on Prairie grasslands but eventually became a victim of habitat loss, overhunting, and predation. In the United States three subspecies of prairie chicken are threatened and an eastern subspecies, the heath hen, became extinct in the 1930s.*

Above: One of the hardiest of all birds, the rock ptarmigan (Lagopus mutus) *is mainly an Arctic species that lives on the rockiest, barest barrens and uplands. While some remain on nesting grounds all winter, others move to lower elevations but never seek cover. To help them walk on snow, these birds grow scales along each toe and feathers on both sides of the feet in winter. Claws also grow longer. Together these changes increase the bearing surface of the feet by 400 percent, reducing the depth the birds sink in the snow by half.*

Opposite: The multicoloured ring-necked pheasant (Phasianus colchicus) *cock, an Asian native, was widely introduced in North America beginning in 1857. It inhabits cultivated farmland, where it pecks at corn, wheat, oats, barley, hay and grasses, and uses hedges for cover and travel lanes. The ring-necked pheasant likes to dust-bathe in the sun. When threatened by a predator such as a hawk, if possible it ducks into a nearby woodchuck burrow.*

Ring-necked pheasant (Phasianus colchicus) *hens gather in groups that are defended by a solitary cock, which closely guards its breeding territory. The cock then abandons the hens and chicks, and sexes usually remain separate except to breed. Some hens lay their eggs in the nests of ducks or other birds. Females protect their offspring by distracting predators from the nest.*

Nicknamed "fool hen," the spruce grouse (Dendragapus canadensis) *is born without an instinctive fear of humans, a trait that has led to its disappearance from places where people have settled. It lives in coniferous forests as far north as the tree line, where it feeds on needles, buds, and berries.*

Perching Birds

Three-fifths or about 5,100 of the world's 8,600 avian species are perching birds. From wrens to ravens, in Canada the passerines comprise about 235 species in 21 families. All but one of Canada's passerine families—tyrant flycatchers—are songbirds. Though the calls of flycatchers could be considered musical, the songs of warblers, thrushes, vireos, finches, larks, bluebirds, and many others are the most highly developed. Among these songbirds is the white-throated sparrow, the unofficial "Canada Bird," known for its distinctive whistling of "Oh sweet Canada, Canada, Canada."

The voices of these woodland singers are controlled by four pairs of syrinx muscles, which make them capable of emitting more elaborate songs than other birds. Used in courtship and mating, bird songs are specific sounds always repeated in the same sequence. Almost invariably, only males sing and the songs

Stashing food for winter is the main occupation of the Clark's nutcracker (Nucifraga columbiana). *Remembering the locations of up to a thousand caches, it may store thirty thousand conifer seeds a year. A hundred seeds can be carried at once in its throat pouch and stored in the ground on south-facing slopes. The cached seeds sustain the birds from winter until midsummer.*

are used as a substitute for battle when warding off adversaries.

Perching birds, particularly the family Corvidae—crows, jays, and magpies—are said to be the most adaptive and intelligent of feathered fauna. Songbirds vary widely in size from kinglets, warblers, and sparrows, measuring ten to fifteen centimetres long, to the sixty-centimetre raven, weighing nearly a kilogram and opening its wings to more than a metre.

Despite their differences, passerines have numerous similarities. All are land birds and all have feet with three toes in front and one behind, specially designed for perching. There are no muscles in the feet, which are controlled by tendons attached to muscles in the legs. When the bird lands, the feet automatically grasp the branch. As long as weight remains on the feet the claws remain wrapped around the branch, allowing the bird to sleep when perched in a tree. Essentially the feet can be used only for grasping and hopping, and most perching birds hop, rather than walk, while on the ground.

The offspring of passerines are hatched naked and blind and rely heavily on their parents. They stimulate the adults to feed them by begging for food and waving their heads with mouths agape. Most nestlings have feeding targets, markings in the mouth where parents place the food. The mouth interiors of birds that nest in well-illuminated areas may be orange or red, while cavity nesters might have white or yellow markings, which can be better seen in poor light.

Some nestlings have nerves that cause the mouth to open when touched by parents; others are stimulated by movement in the nest or by high-pitched calls from the adults. There's no pecking order at dinner time: the closest mouth gets the meal. Nestlings, however, can't swallow a second piece of food immediately after the first, so an unswallowed morsel is retrieved and shoved into the mouth of another hungry offspring.

Some Canadian songbirds have long migrations while others stay in nesting areas throughout the year. Tree swallows that breed in Canada may winter in

As these young horned larks (Eremophila alpestris) *mature, they will develop short, black feathered horns that protrude from the back of the head like miniature antlers. This species thrives in treeless terrain—Arctic tundra, prairies, open sea coasts, and farms. Though farming sometimes destroys ground nests, populations of horned larks are expanding in parts of North America where clearing of land has created more habitat.*

Cuba; bobolinks might migrate to Argentina or Brazil, and Clark's nutcrackers may fly to New Mexico or Baja California. Some chickadees, jays, crows, and ravens rarely migrate, managing to find a year-round food supply close to home, often with the help of humans. The urge to migrate varies not only with the species, but with individual birds. Some migrate one year and not the next.

For most of the year the fat content in a bird's body is about 5 percent. But as changes in daylight hours and seasonal food sources occur, some birds prepare to migrate by increasing their body fat to 40 or 50 percent.

Migration routes are thought to be ancestral, and while annual southbound and northbound routes may differ, nearly all birds follow the same courses each year. They navigate by natural landmarks: daytime migrators such as swallows, robins, or blackbirds use the sun to determine direction; night fliers like tanagers, orioles, flycatchers, warblers, sparrows, or vireos are guided by the stars. Wind, scent, and a built-in magnetic sense are also navigational aids to migrating birds. Skyscrapers, however, are navigational hazards. When night-flying migrators are forced by clouds to fly low, they become confused and attracted by lights from tall buildings: tens of thousands of dead birds have been shovelled from streets below skyscrapers after overcast nights.

In Canada the Kirtland's warbler is the only endangered perching bird. There may be fewer than four hundred left in the world.

Described by Canadian ornithologist W. Earl Godfrey as "the personification of cheerfulness and good nature," the black-capped chickadee (Parus atricapillus) *delights many a birder across all of southern Canada. It remains here year-round, often relying on help from humans to survive cold Canadian winters. Canada even has a children's magazine named after this engaging bird.*

American crows (Corvus brachyrhynchos) *gather under the watchful eyes of a timber wolf. Crows often follow hunting wolves in the hope of sharing a kill. Highly adaptable crows are found nearly any place in Canada where food is sufficient. Like other members of the family Corvidae, they are regarded as extremely intelligent, capable of mimicking many other animals.*

The antagonistic eastern kingbird (Tyrannus tyrannus) *is intolerant of trespassers in its territory. It fearlessly attacks crows, hawks, vultures, and other much larger birds and has even been known to try and ward off low-flying aircraft. A member of the flycatcher family, it occurs across most of southern Canada.*

Above: Canada's largest perching bird, the common raven (Corvus corax) *is found in virtually every part of the country except the extreme southern Prairies and Niagara Peninsula. This is perhaps the brightest of Canadian birds, a master of mimicry that engages in complex play and works in teams to steal meals from other animals. Ravens figure prominently in the art and folklore of native Indians across Canada.*

Previous page: The crimson crown and pinkish back and breast are the most distinctive features of the purple finch (Carpodacus purpureus), *a common sight across most of southern Canada. It builds nests in coniferous forests, occasionally orchards. In some areas its numbers fluctuate as it competes for nesting and foraging space with house finches.*

The casual whistles of the white-crowned sparrow (Zonotrichia leucophrys) *are heard across much of Canada, particularly in the north. Experiments with this species suggest their songs are learned during the first seven weeks of life. At about twenty-one weeks they begin to practise the tunes they learned as youngsters. By the twenty-eighth week, in time to attract breeding partners, their songs are strikingly similar to those of their parents and other white-crowned sparrows.*

Above: The appealing voice of the song sparrow (Melospiza melodia) *is one of the earliest spring birdsongs in Canada. One of the most common and pervasive sparrows across southern Canada, it not only arrives sooner than most spring migrants, but continues to sing after most species have stopped in midsummer.*

Opposite: A bird of Canada's most northerly climes, the snow bunting (Plectrophenax nivalis) *is a true lover of winter. After nesting in rock crevices and cliffs of the Arctic, it migrates to much of snow-covered southern Canada, where it feeds in large flocks on seeds from weeds that protrude through the snow. It leaves for its northern mating grounds before the snow has melted from its wintering areas.*

The song of the white-throated sparrow (Zonotrichia albicollis)*—*
"Oh sweet Canada, Canada, Canada"—has earned it the nickname
"Canada bird." Its enchanting, repetitive whistle—two melodic
notes followed by a phrase of three notes—is heard across most
of the country except the far north and the Pacific coast.

With a few exceptions, only residents of southeastern Ontario and Quebec enjoy the sight of bright red northern cardinals (Cardinalis cardinalis) *in Canada. Fortunately, it appears that its range is gradually expanding. Breaking with the songbird tradition that only males sing, the females of this species sing after males have established territories and before nesting begins.*

Above: Across Canada the American robin (Turdus migratorius) *is welcomed as the first sign of spring. Its appearance on neighbourhood lawns brings a feeling of optimism for winter-weary Canadians. Contrary to popular belief, experiments have proven that robins do not catch worms by cocking their heads to one side and listening to the worm's movements underground: they hunt by sight.*

Previous page: The only crested jay in the west, the Steller's jay (Cyanocitta stelleri) *inhabits southern Canada from the Rockies to the Pacific. On the southwest coast its favourite food is Garry oak acorns, which it stores in hundreds of caches. These raucous birds are thought to be one of the major distributors of Garry oak trees. In 1987 British Columbia adopted the Steller's jay as the provincial bird.*

Bohemian waxwings (Bombycilla garrulus) *invariably show up unexpectedly. They do not establish territories, but travel about in flocks, often with robins, looking for fruit, berries, insects, and tree sap. Their feathers secrete a waxy substance that collects on the tips of their wings like bright red sealing wax. The purpose of the wax is unknown.*

Above: Like its cousins the Steller's and blue jays, the grey jay (Perisoreus canadensis) *is an audacious freebooter that shamelessly pilfers camps and picnic tables. Occurring across most of Canada, this bold bird shows little fear of humans. It's a particularly hardy species, often nesting when snow is still deep on the ground.*

Opposite: Only southern Canadian residents east of Alberta have the opportunity to observe eastern bluebirds (Sialia sialis) *and listen to their pleasant warblings. But numbers here are declining by as much as 90 percent in some areas, partly because of competition with starlings, house sparrows, and other cavity nesters. Artificial nest boxes are bringing back the eastern bluebird to some areas, but it remains listed as "vulnerable."*

The cheery, sonorous call of the red-winged blackbird (Agelaius phoeniceus) *is heard across all of southern Canada. The males of this species seem constantly on the move, displaying their red-orange shoulder crests as they flit from branch to branch. They build loosely woven nests over water in dense marsh grass. Occasionally the youngsters fall out of the nests, so young blackbirds learn to swim within their first week.*

Like a small parrot, the white-winged crossbill (Loxia leucoptera) *climbs in trees, using its beak as a third foot. A resident of coniferous forests, the tips of its mandibles cross one another. The bird inserts the beak tip into a cone, forcing the scales apart and extracting the seed with its tongue. This bird is found in most of Canada except the high Arctic and southern Prairies.*

Above: A northern nester, the common redpoll (Carduelis flammea) *is better able to survive harsh Arctic conditions than other songbirds. In cold winds it finds shelter in coniferous forests and stands motionless, with feathers puffed up, to reduce loss of heat. During winter, redpolls are a common sight in much of southern Canada.*

Opposite: One of Canada's most widely distributed birds, the tree swallow (Tachycineta bicolor) *nests in cavities or man-made boxes. A bug-eater, it hunts on the wing over wet ground or lakes, ponds, sloughs, and streams, occasionally splashing on the surface as it snatches its prey.*

Like other members of the family Corvidae, the blue jay (Cyanocitta cristata) *is as curious as it is mischievous. A resident of southern Canada east of the Rockies, it is one of the most common year-round birds of the east. It is a notorious thief, raiding nests of small birds for eggs and offspring. By perfectly imitating the call of a sparrow hawk, the blue jay scares smaller birds into taking cover in the bush.*

Above: A native of streams and rivers in British Columbia, southwestern Alberta and the western Yukon, the American dipper (Cinclus mexicanus) *is a unique semiaquatic bird, the only member of the family Cinclidae in Canada. With scales that cover its nostrils when submerged and oil glands much larger than those of other passerines, the dipper catches bugs and small fish to depths of six metres.*

Overleaf: A chestnut-sided warbler (Dendroica pensylvanica) *feeds its hungry chick. In the early 1800s these birds were scarce in much of North America, but regeneration of deciduous forests and bushy thickets provided new nesting habitat. Now this bird breeds across southern Canada east of Alberta.*

PHOTO CREDITS

Annie Griffiths Belt/First Light p. 117
Tim Christie p. 140
Ralph Clevenger/First Light p. 83
Tim Fitzharris p. 100
Dawn Goss/First Light p. 146
Chris Harris p. vi
Stephen Homer/First Light pp. 31, 36
Victoria Hurst/First Light pp. 48, 115
Thomas Kitchin/First Light pp. 3, 9, 12, 15, 20, 41, 52, 55, 57, 61, 63, 64, 68, 71, 77, 78, 80, 82, 84, 86, 87, 88, 89, 92, 94, 96, 103, 104, 109, 118, 119, 120, 138, 144
Todd Korol/First Light p. 98
Robert Lankinen/First Light pp. 14, 17, 26, 32, 33, 42, 47, 49, 58, 76, 90, 108, 111, 112, 116, 121, 126, 130, 133, 141, 142, 145, 150
Scott Leslie/First Light pp. 28, 93, 99, 136, 137, 148
Alan Marsh/First Light pp. 44, 46
Peter McLeod/First Light pp. 23, 128, 135
Brian Milne/First Light pp. 4, 16, 19, 24, 25, 30, 62, 67, 125, 129, 134, 149
Pat Morrow/First Light p. 114
Bruce Obee p. 29
G. Petersen/First Light p. 18
Charles Philip/First Light pp. 10, 34
Duane Sept pp. i, 50, 56, 60, 66, 132, 147
A.E. Sirulnikoff/First Light p. 51
Donald Standfield/First Light pp. 35, 54
John Sylvester/First Light pp. 122, 143
Wayne Wegner/First Light pp. 22, 37
Darwin Wiggett/First Light pp. 38, 72, 74, 81, 95, 106, 110, 113
Dale Wilson/First Light p. 6

INDEX

Photographs in italics